Natural Disasters

Howling Hurricanes

Julie Richards

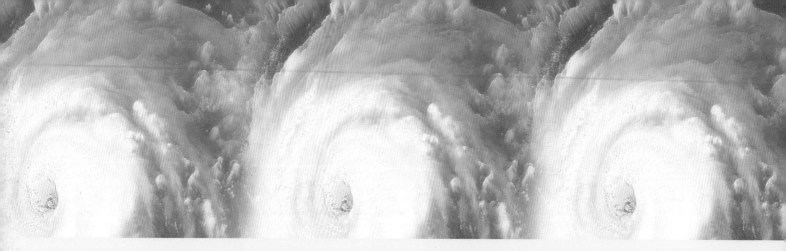

This edition first published in 2002 in the United States of America by Chelsea House Publishers, a subsidiary of Haights Cross Communications.

Chelsea House Publishers
1974 Sproul Road, Suite 400
Broomall, PA 19008-0914

The Chelsea House world wide web address is www.chelseahouse.com

Library of Congress Cataloging-in-Publication Data Applied for.

ISBN 0-7910-6584-7

First published in 2000 by
Macmillan Education Australia Pty Ltd
627 Chapel Street, South Yarra, Australia, 3141

Copyright © Julie Richards 2001

Edited by Sally Woollett
Text design by Polar Design Pty Ltd
Cover design by Polar Design Pty Ltd
Illustrations and maps by Pat Kermode, Purple Rabbit Productions
Printed in Hong Kong

Acknowledgements
The author and the publisher are grateful to the following for permission to reproduce copyright material:

Cover photograph: Satellite image of hurricane, courtesy of PhotoDisc.

AAP Image/Bureau of Meteorology, pp. 8 (top), 18; Australian Picture Library/CORBIS, pp. 26, 28; Australian Picture Library/John Carnemolla, p. 27; Bureau of Meteorology, pp. 11, 16, 17 (bottom); CSIRO, p. 19; Department of Foreign Affairs and Trade, pp. 29, 30; Keesler AFB, United States Air Force, pp. 9 (top), 22 (bottom); PhotoDisc, pp. 2–4, 8 (bottom), 14, 15, 17 (top), 23, 30, 31; Photolibrary.com, pp. 12, 13; Reuters, p. 24–25; Scott A. Dommin, pp. 5, 9 (bottom), 20–21, 22 (top).

While every care has been taken to trace and acknowledge copyright the publishers tender their apologies for any accidental infringement where copyright has proved untraceable. Where the attempt has been unsuccessful, the publisher welcomes information that would redress the situation.

Contents

A hurricane is coming

Gathering clouds are towering up into the sky like a thick, black wall.

The **wind whistles** through overhead power lines and between buildings.

Raindrops start falling. **Waves** are coming further up the beach than ever before.

Now **the wind is so strong** that it blows you over and away! It roars past your body so fast that you cannot breathe in.

Rain pounds on walls and windows, pouring into homes as the wind peels away roofs.

HURRICANES are a natural part of the Earth's weather. Scientists believe that hurricanes help to balance the Earth's temperatures by getting rid of extra heat. In different parts of the world, many farmers depend on these massive storms to bring enough rain for growing crops. However, a hurricane can roar into a place where lots of people live. It might kill and injure thousands of people and animals and cause a lot of damage. This is when a hurricane becomes a natural disaster.

Suddenly, the ocean seems to pull itself up into **one huge wave**. As the wave crashes onto the sand it swallows the whole beach and pours into the streets! Houses, cars and furniture are swept away on the tide.

The hurricane has arrived.

What is a hurricane?

A hurricane is a swirling mass of clouds that spins around a calm, clear center called an **eye**.

Hurricanes are the largest, most powerful storms in the world.

HURRICANE FACTS

A hurricane can:

☑ grow to 1,600 kilometers (1,000 miles) across

☑ reach higher into the sky than passenger aircraft can fly

☑ have winds that almost reach the same speed as Formula One racing cars

☑ hold the same amount of water as a large dam

☑ destroy an area 500 kilometers (300 miles) across

☑ have winds that make the same amount of noise as the engines of a jet aircraft on take-off

☑ release as much energy every second as the nuclear bomb dropped on the Japanese city of Hiroshima in 1945.

The dark space in the center of the hurricane's white swirling clouds is the eye.

Where do hurricanes form?

All hurricanes form over the ocean. There are three oceans in the world where hurricanes develop. Hurricanes are also called typhoons or cyclones. The different names are used to tell us which ocean the storms are forming in.

➤ Hurricanes form in the Atlantic Ocean.

➤ Typhoons form in the North and South Pacific Ocean.

➤ Cyclones form in the Indian Ocean.

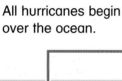

All hurricanes begin over the ocean.

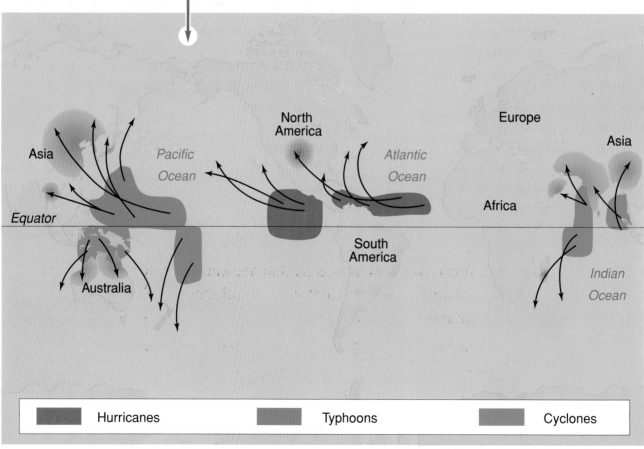

■ Hurricanes	■ Typhoons	■ Cyclones

How does a hurricane happen?

Movement in the atmosphere

The Earth is wrapped in a blanket of gases called the **atmosphere.** The atmosphere is like an enormous weather machine—all of the world's weather is made inside it. The Sun is like a giant engine driving the weather machine. The Sun's energy drives air and water around our planet by creating the clouds and the **wind** as it warms the Earth.

The movement of air and water around the Earth.

GUESS WHAT?

The Earth's atmosphere is 500 kilometers (300 miles) thick.

3 As the warm air rises, cooler air rushes in to fill the space left behind. This movement of air is what we call wind.

2 Droplets of water change into an invisible gas called water vapor and rise into the air above the oceans, lakes and rivers.

4 The invisible water vapor rises into the cooler air near the top of the atmosphere. As the air becomes colder, the water vapor turns back into water droplets. We see them as clouds.

1 When the Sun shines onto the Earth it warms the land and the oceans. The air above the ground warms and begins to rise.

Warm oceans

In warmer places, the 'weather machine' can be very powerful, because there is more energy from the Sun to drive it. This means that hurricanes form over seas or oceans that are warm. The warmest oceans are found near the Equator, an imaginary line that runs around the middle of the Earth.

Because the air at the Equator is very warm, it rises very fast, sucking **water vapor** up into the atmosphere like a giant pump. All this rising air forms clouds, which can then build into thunderstorms. When these thunderstorms join together, we have the beginnings of a hurricane.

GUESS WHAT?

Bogor, in Indonesia, has more thunderstorm days than anywhere else in the world. Out of 365 days, 322 are thunderstorm days.

Life cycle of a hurricane

It takes time for a storm to develop into a hurricane strong enough to cause destruction. The average life of a hurricane is five to ten days. As they develop, most storms will follow a pattern like this:

1 A thunderstorm forms near the Equator. It draws in so much air and moisture it keeps growing larger and larger. As it grows, it joins with other thunderstorms nearby. This is what meteorologists call a tropical depression.

2 When the air moving inside this storm reaches a speed of 63 kilometers (39 miles) per hour it becomes a tropical storm. The storm does not weaken. Instead, an eye begins to form in the center of the clouds and the wind spins even more quickly.

3 When the wind speed inside the storm increases to 119 kilometers (74 miles) per hour and begins to spiral tightly around the eye, it becomes a hurricane.

4 As well as spinning, the circle of cloud, wind and rain is also moving forwards. While there is plenty of warm, moist air to feed it, the hurricane will grow stronger as it travels across the ocean.

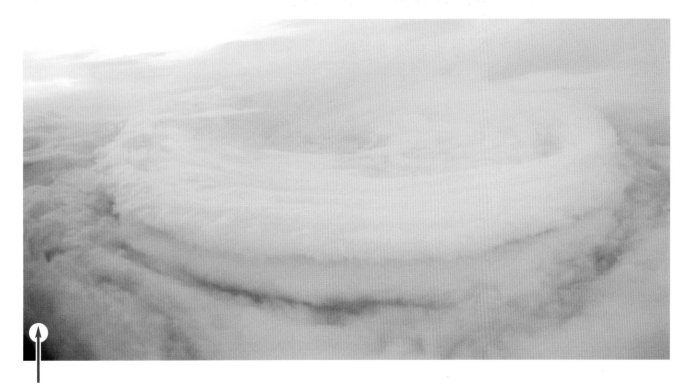

5 Finally, the hurricane will reach the coast and cross onto dry land. This is called **landfall**. Away from the warm, moist air of the ocean, the hurricane will begin to weaken and, soon after, it will disappear.

The eye

When the eye passes over, the hurricane's vicious winds and hammering rain will suddenly stop. Even the sun might break through! The eye is always calm and clear like this. It can fool you into thinking the storm is over. The clouds circling the eye are called the **eyewall**. The eyewall contains the fastest wind and the heaviest rain.

The eye may be between 5 and 105 kilometers (3 and 65 miles) across. It divides the hurricane into two parts. It is only the first part of the storm that is over. Take cover! The second part is nearly here and will be every bit as fierce and frightening as the first part.

Inside a hurricane

The spinning movement of the Earth makes the storm clouds and the winds spin in the same direction. In the Northern Hemisphere hurricanes and typhoons spiral in a **counterclockwise** direction, while in the Southern Hemisphere cyclones spiral in a **clockwise** direction.

Air movement inside a hurricane.

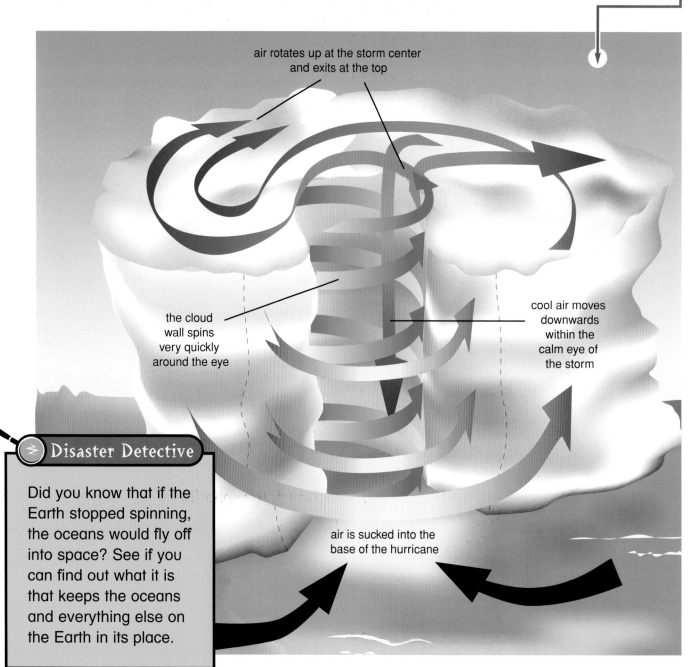

air rotates up at the storm center and exits at the top

the cloud wall spins very quickly around the eye

cool air moves downwards within the calm eye of the storm

air is sucked into the base of the hurricane

Disaster Detective

Did you know that if the Earth stopped spinning, the oceans would fly off into space? See if you can find out what it is that keeps the oceans and everything else on the Earth in its place.

Which way will it go?

Winds blowing outside the hurricane will guide it across the ocean, much like a wheel steers a car. Hurricanes do not travel in perfectly straight lines. Most wander backwards and forwards before striking the coast. Sometimes, one will come very close, but swing out to sea again at the last moment. Another, not expected to cross the coast, may suddenly do so. You can see how difficult it must be for meteorologists to know which people to warn.

On Christmas Eve, 1974, cyclone Tracy swerved around Melville Island, bringing the people of Darwin, Australia an unwelcome early Christmas present.

Read All About It!

The world's longest running storm

Hurricane John lasted for 29 days and travelled 8,000 kilometers (5,000 miles) across the ocean. The storm travelled so far that it crossed over into typhoon territory, and its name had to be changed to typhoon John. When the storm wandered back again it was renamed hurricane John!

How fast will it move?

Although the spiralling winds inside the hurricane may reach incredible speeds, the hurricane itself moves quite slowly across the ocean: its top speed is about 50 kilometers (30 miles) per hour.

What kinds of damage can a hurricane do?

Hurricanes cause damage in three different ways.

Storm surge

The sucking power of a hurricane is so great that it can lift the sea beneath it. When this happens, the hurricane pushes the sea ahead of it. As it nears the shore, the water is slowed down by the land and piles up into gigantic waves of six meters (20 feet) or more, before smashing its way inland. Meteorologists call this a **storm surge**. Nine out of ten people killed in a hurricane are drowned by the storm surge.

Storm surges can be 65 to 80 kilometers (40 to 50 miles) across. This means they can flood homes and damage beaches outside the hurricane's main danger zone.

STORM SURGE FACTS

☑ Hurricane Andrew caused one of the largest storm surges on record: seven meters (23 feet). It destroyed 75,000 fully grown trees.

☑ Hurricane Hazel sliced 272 beachfront cottages off their foundations before blasting through orchards and sending millions of apples bouncing from their trees.

☑ Hurricane Hugo cut an island in two and swept a fishing boat eight kilometers (five miles) inland, leaving it in a forest!

A hurricane and storm surge have destroyed these homes.

GUESS WHAT?

During hurricane Camille the storm surge flushed alligators and hundreds of poisonous snakes out of creeks and swamps. More than 40 people died from snake bites.

In 1991, cyclone 2B, with a six meter (20 foot) storm surge, flooded the country of Bangladesh, near India. More than 140,000 people were drowned and 5,000 fishermen did not return from the sea that day. One and a half million homes were washed away. Millions of tons of crops were ruined and one million cattle drowned. This was nothing compared to the 1970 cyclone. That storm claimed more than 500,000 lives and was one of the worst natural disasters of the twentieth century.

A storm surge can arrive in minutes, hours or days. Sometimes it will arrive a week before the hurricane itself does. The shape of the coastline can be changed forever when huge, surging seas wash away cliffs and beaches or move small islands.

Wind

Many hurricanes have winds blowing at 240 kilometers (150 miles) per hour. Such high winds can blow down buildings, wrench trees out of the ground and flip cars and small aircraft. They can also whip up enormous waves far out at sea, threatening even large ships.

However, there is an even greater danger: these winds are so ferocious that they can pick up large objects and hurl them through the air. Even small objects travelling at high speed can kill if they hit someone. The wind can also rip down overhead power lines, which can electrocute people who touch them.

Small hurricanes can be just as destructive as large ones, especially if they happen in an area where a lot of people live. The winds of cyclone Tracy only spread out 50 kilometers (30 miles) from the center of the storm, but that was enough to demolish nearly every building in Darwin, Australia.

TRY THIS

1 Hold a sheet of thin paper so that one edge is level with your lower lip and the rest of the sheet curves downward.

2 Blow across the sheet. What happens? This is how a hurricane lifts the roof from a building when it blows across it.

Hurricanes are strong enough to lift and throw even large, heavy objects like this airplane.

Rain

Moving mountains

Even without a storm surge, the rain from a hurricane causes terrible flooding. It can loosen rocks and soil on hillsides and mountain slopes. When the rain mixes with the soil it becomes mud and washes down onto everything below. This is called a landslide or a mudslide.

GUESS WHAT?

When a hurricane killed 100,000 cattle, the job of collecting and burying the rotting carcasses seemed so enormous that everyone was afraid disease would spread before the job could be finished. However, flocks of vultures landed and in a few days only the clean bones of the animals were left.

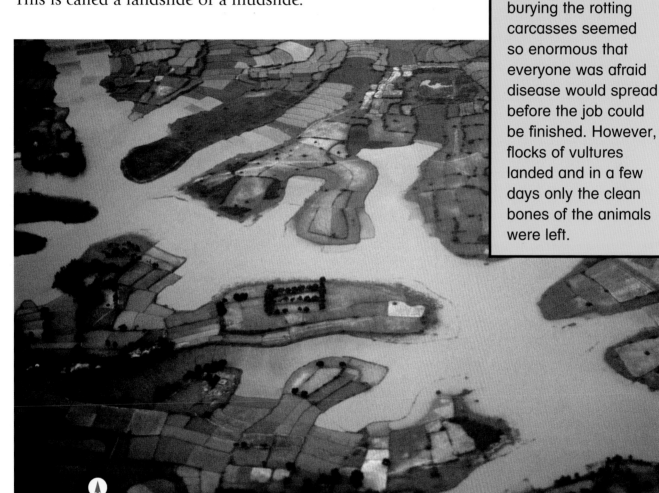

A scene like this can often mean starvation and disease because crops and drinking wells are ruined by salt water.

Floods and disease

Flooding may cause drains and sewers to burst. Dirty waste water can overflow into dams that hold clean drinking water. The bodies of drowned people and animals become mixed up with all sorts of rubbish and wreckage left behind by the hurricane. When this happens, all the bodies and animal carcasses must be collected and then buried or burned. If they begin to decay, serious diseases can spread very quickly.

Forecasting and measuring hurricanes

Hurricane season

Hurricanes form at certain times of the year. This is called hurricane season. During this time, weather scientists are always on the lookout for the type of weather that can start a hurricane brewing.

Lots of warm, moist air is needed to start a hurricane. This means that hurricane season is usually during summer and early autumn.

➤ Hurricanes form from June to November in the **Northern Hemisphere**.

➤ Typhoons form from April to December in the Northern Hemisphere.

➤ Cyclones form from December to March in the **Southern Hemisphere**.

Weather scientists

Weather scientists are called **meteorologists**. One of the jobs of a meteorologist is to look out for hurricanes and warn people who might be in danger.

The information meteorologists give us about the weather is called a forecast or prediction. However, the weather does not always do what meteorologists expect it to.

Although many people are threatened by hurricanes, these storms do not happen every day, nor do they happen in all parts of the world.

Meteorologists try to let us know what sort of weather is heading our way.

GUESS WHAT?

The first weather map was published in 1851 and displayed at the Great Exhibition in London, England. The first weather forecast to be published appeared in the *Times* newspaper in London, during the 1860s.

Hurricane tracking equipment

Hurricane season is here again and all the ingredients for a hurricane are coming together. Using special equipment, meteorologists will keep watching this weather.

Satellites

Just outside the Earth's atmosphere, **satellites** collect information about the world's weather. Satellites have instruments that can measure the air temperature in different parts of the world. The information collected is then sent back to Earth. Computers can change this information into pictures and maps for meteorologists to use when making forecasts.

Radar

Radar sends out a signal that bounces off the hailstones and raindrops inside clouds. These signals send information about the position, movement and size of cloud formations back to the weather station, where meteorologists can see it on a special screen.

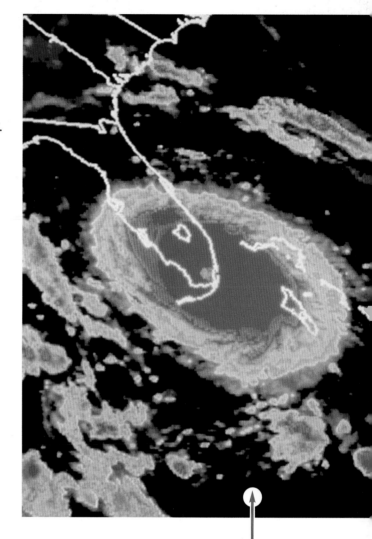

A satellite image of a developing hurricane.

Radiosondes

A package of weather instruments attached to a weather balloon is called a **radiosonde**. All over the world, meteorologists release radiosondes twice each day. Radiosondes report on weather conditions high up in the atmosphere and send back information that is helpful for forecasting.

Meteorologist launching a radiosonde.

Hurricane alert!

Imagine that satellite pictures of your part of the world show small thunderstorms joining to form one vast storm. Its winds are blowing at 120 kilometers (75 miles) per hour. The rain falling from its dense, black clouds is so heavy that it looks like a solid sheet. In the center, the eye of the storm is clearly seen. Meteorologists decide to declare this storm a hurricane. Over the next few days, for the life of this storm, meteorologists are going to be extremely busy.

Naming hurricanes

The first thing the meteorologists do is give the new hurricane a name. Because another hurricane may form in the same area at the same time, it is important to make sure that everyone understands which hurricane the meteorologists are talking about.

In each part of the world where these storms happen, a different list of names is made available to meteorologists. The lists work alphabetically. The first hurricane of the season is given a name beginning with the letter A, the second with the letter B, and so on. The names repeat every six years.

When a storm causes terrible damage and loss of life, its name is removed from the list. Andrew, Gilbert and Tracy are just three of the names that will never be used again.

> **GUESS WHAT?**
>
> In the early 1900s, an Australian weather forecaster called Clement Wragge was the first person to name hurricanes. He named them after people he did not like!

This satellite picture shows tropical cyclones Steve and Norman approaching the coast of Australia in March, 2000.

> **Disaster Detective**
>
> Perhaps you can find out if there has ever been a hurricane, typhoon or cyclone with the same name as yours.

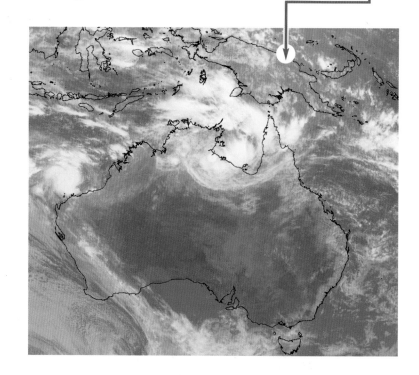

Gust probes

A gust **probe** will measure the dangerous wind gusts that blow through the lower part of the hurricane. Gust probes are trailed from an aircraft that can fly over the top of the storm.

Saffir–Simpson Hurricane Scale

The power of a hurricane is measured using a special scale called the Saffir–Simpson Hurricane Scale. When meteorologists know how fast the hurricane's winds are blowing, they will try and predict how high the storm surge will be. The hurricane will then be given a rating from Category 1 to Category 5.

Weather aircraft have measuring instruments in their long nose probes so that the instruments are not disturbed by the air rushing around the engines.

This scale helps everyone to prepare for the hurricane by showing them the type of damage to expect when it arrives.

Category 1 (weak)

Category 2 (moderate)

Category 3 (strong)

Category 4 (very strong)

Category 5 (devastating)

Hurricane hunters

Meteorologists also get important information about the new hurricane from a group of very brave pilots called hurricane hunters.

Hurricane hunters fly special aircraft fitted with lots of weather instruments. They fly right through the thick clouds of the eyewall. This is the most dangerous part of the hurricane, so it is always a very bumpy ride!

Flying with the hurricane hunters

The hurricane hunters' aircraft has a crew of six.

➤ Two pilots are needed to keep the big Hercules aircraft safe and steady as it flies through heavy rain, lightning and gusting winds.

➤ The navigator will track the storm on radar. The pilots rely on the navigator to warn them of approaching dangerous weather and how to safely avoid it.

➤ The flight engineer checks the fuel levels and makes sure that the big turboprop engines are running smoothly.

➤ The weather officer will keep the meteorologists on the ground up to date on the storm. Every 30 seconds, the weather officer's computer sends information to the meteorologists.

➤ The **dropsonde** operator is responsible for loading the dropsonde into its launch tube. The dropsonde looks very much like a torpedo. It contains more weather-measuring instruments.

The first job for the hurricane hunters is to find the eye of the storm. This is known as getting a **center fix**. They can then measure the wind speed and tell the meteorologists exactly where the center of the storm is. Each journey the hurricane hunters make into the storm is called a **cut**.

> **GUESS WHAT?**
>
> Ships survive hurricanes at sea by sailing fast enough to stay inside the eye. When this happens, flocks of sea birds often land on the ships and ride out the storm with them!

The Hercules WC-130 is one type of aircraft flown by the hurricane hunters. It flies through different parts of the hurricane to gather as much information as possible about the storm.

Using the information

The information gathered by the hurricane hunters is important to meteorologists because it helps them to:

➤ forecast when and where the hurricane will cross the coast

➤ decide whether the hurricane is weakening or growing stronger

➤ work out how high the storm surge will be

➤ give the hurricane a rating on the Saffir–Simpson Hurricane Scale

➤ decide which people need to be warned or moved out of the hurricane's way.

What the hurricane hunters see

The navigator has spotted a thick band of violent thunderstorms on his radar. He warns the pilot and the aircraft swerves around them. It is only eight kilometers (five miles) to the eyewall. Rain begins to pelt the windows. It is getting very dark; the aircraft shakes and rolls as the wind grabs it. Below, the sea is **churning**. The pilot orders everyone to strap themselves into their seats. There is a burst of lightning; the aircraft drops suddenly...

It seems like forever, but then it happens! The sky brightens and the rain stops. One last bump and the aircraft flies smoothly again. A solid wall of clouds circles the aircraft. The hurricane hunters have made it safely through to the eye!

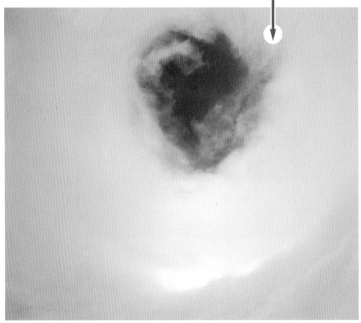

Looking up through the eye at a clear, sunny sky.

The dropsonde operator loads the dropsonde into its launch tube.

The most important information of all

Once the hurricane hunters have reached the eye, the operator will launch the dropsonde into the ocean below. Just before it splashes into the water, it will send back the most important piece of information: a measurement that will tell meteorologists if the hurricane is becoming weaker or growing stronger.

Preparing for a hurricane

As a hurricane approaches the coast, people prepare for its arrival. At nearby harbors and marinas two red flags with black squares in their centers will be hoisted up flag poles. These are marine hurricane warning flags. They let sailors know that it is too dangerous to leave the harbor.

Meteorologists use **super computers** and the information sent to them by the hurricane hunters to track the hurricane. They try to work out who will be in danger so that these people can be warned to leave before the hurricane blows in. Moving people away from danger is called evacuation. When large numbers of people have to be moved at the same time it takes a lot of planning. A special timetable is drawn up to make it easier.

Hurricane warning signs tell people that a hurricane is approaching.

Hurricane
颶風

Read All About It!

In 1938, just before a hurricane hit Long Island on the east coast of the USA, a man received a weather forecasting instrument called a barometer through the mail. The needle on the instrument pointed to the word 'hurricane'. When the needle would not move, the man thought it was broken. Angry, he immediately sat down and wrote a letter complaining to the manufacturer. When he returned from posting the letter, he found that his house had been destroyed by a hurricane.

Hurricane watch

Thirty-six hours before the hurricane is expected to arrive, a hurricane watch is broadcast on radio and television. There are lots of jobs to be done during this time before the evacuation begins.

Anything that could become a dangerous missile in high winds, such as garden furniture and bikes, must be put away. Even small buildings such as sheds need to be anchored to the ground. Nailing shutters to the windows will help prevent them from shattering. Animals can be killed, injured or lost during a hurricane just as easily as people can, so pets need to be rounded up and cared for too.

People pack their cars with everything they will need while they are away. Even if they are not being evacuated, they will still need supplies to see them through the storm. It may be some time before the electricity is back on, telephones work or clean drinking water is available.

Hurricane warning

Twenty-four hours before the hurricane arrives, a hurricane warning is broadcast. Television weather maps will show the danger zone in red. Areas just outside the danger zone are colored yellow. These yellow areas may experience flooding, even though they may not be directly in the path of the hurricane. Storm surge waves can still reach parts of the coastline outside the hurricane danger zone.

Maps showing evacuation routes are advertised so that everyone knows the safest and quickest roads to travel.

People who live in areas where hurricanes are common know how to prepare during a hurricane watch.

Evacuation

The people in the greatest danger are those who live in mobile homes. They will be moved into community shelters. Oil drilling rigs at sea may be in the path of the hurricane. Helicopters will airlift the workers from the drilling platforms to safety. Most of the people evacuated will stay with family or friends who live far enough away from where the hurricane is expected to cross the coast.

Evacuation helps people avoid injury and loss of life. However, some people do not want to leave. Some may need to be rescued from their wrecked cars or the ruins of their homes after the hurricane has passed. During the 1950s and 1960s, instead of evacuating, people actually held hurricane parties to celebrate the arrival of a storm. In 1969, one celebration ended suddenly when hurricane Camille gate-crashed the party. Only one party guest survived.

HURRICANE FACTS

A hurricane supply list

- ☑ bottled water
- ☑ canned food
- ☑ blankets
- ☑ first aid kit
- ☑ transistor radio
- ☑ flashlights

What might happen in a developing country?

Bangladesh is one of the world's poorest and most crowded countries. Even though community shelters have been built in Bangladesh, there are still a lot of people who do not have televisions and radios. Many of them never hear or see the warnings that could save their lives.

In the event of a hurricane, countries such as Bangladesh need help from other countries to care for the survivors, to clean up the wreckage and to rebuild towns and cities.

Can hurricanes be prevented?

Hurricanes cannot be stopped. In the past, scientists tried to break up hurricanes by dropping special chemical crystals into the clouds. It was not very successful.

Other suggestions have included exploding a nuclear bomb inside the hurricane, spreading plastic sheets over the ocean or tipping ice into it to prevent it from heating up. This may sound very silly, but it does show how much these storms frighten people and how much damage they cause each year.

The best way for people to minimize the damage done by hurricanes is to prepare their homes correctly during a hurricane watch and to construct buildings that are able to withstand them.

The cross-shaped girders on this building help strengthen and protect its enormous glass walls against high winds.

GUESS WHAT?

The Red Spot on the planet Jupiter is really the top of a great spiralling hurricane. This storm is larger than the Earth itself and has been raging for more than 300 years!

Pollution of the atmosphere prevents some of the Sun's heat from returning to space.

⧐ Disaster Detective

Nearly half of Bangladesh is less than one meter (three feet) above sea level. If the sea keeps getting higher, half of the country could be underwater by 2030. Perhaps you could find out if there are other places in the world facing the same danger.

Hurricanes and global warming

Certain gases from factories and car exhausts stop some of the Sun's heat from returning to space. This trapped heat makes the temperature of the Earth rise. This is called global warming. Global warming may cause more hurricanes if the air above the oceans becomes any hotter.

Warmer temperatures have caused ice at the North and South poles to melt more quickly. This makes the sea higher. Higher sea levels may make it possible for hurricanes to travel further inland than ever before.

Global warming and its effects can be reduced by limiting the amount of energy we use and by slowing the rate at which forests are being cleared.

After a hurricane

Hurricanes make a real mess. They leave behind filthy flood waters, mud and all sorts of rubbish. There are many things that need to be done.

➤ People trapped in the wreckage need to be rescued and given medical treatment.

➤ People whose homes have been destroyed need food, water, shelter and clothing.

➤ Cleaning up quickly is very important to help prevent the outbreak of disease.

➤ Buildings need to be demolished or made safe.

➤ Electricity and fresh water must be reconnected.

➤ Power lines, gas pipes and drains need to be checked.

➤ Stray animals need to be rounded up.

It can take months or even years to rebuild towns or cities damaged by hurricanes. If the damage is too great the town may be abandoned.

One of the first jobs after a hurricane is to provide people with shelter, food and clothing.

Cyclone Tracy caused so much damage that Darwin had to be completely rebuilt.

On December 24, 1974, cyclone Tracy brought the people of Darwin in northern Australia an unwanted early Christmas present. Although only 50 bodies were found, nearly every building in Darwin was destroyed or damaged beyond repair. The devastation to Darwin was so great that it was impossible for anyone to live there. Twenty thousand people were airlifted to other parts of Australia. It was decided that the new town must be built to withstand another cyclone like Tracy.

GUESS WHAT?

Hainan province in China was struck by three typhoons in 11 days.

Record-breaking hurricanes

The most costly hurricane in recent history

Hurricane Andrew

In 1992, hurricane Andrew ripped a 40 kilometer (25 mile) wide track of destruction through the Bahamas and across Florida and Louisiana in the USA. The damage bill was 25 billion US dollars. Hurricane Andrew only spent four hours in Florida, but it left 80,000 homes in ruins, and a further 55,000 were badly damaged.

The winds of some hurricanes are strong enough to bend metal power poles.

The hurricanes with the fastest winds

Hurricanes Andrew and Gilbert

Both hurricanes had winds gusting up to 320 kilometers (200 miles) per hour.

The hurricane that developed the fastest

Typhoon Forrest

Typhoon Forrest took less than 24 hours to develop into a mature storm.

The hurricane with the largest wind spread

Typhoon Tip

Typhoon Tip's winds extended 1,100 kilometers (682 miles) out from the centre of the storm.

The highest recorded storm surge

Cyclones Nachon and Mahina

In 1899, a storm surge in Bathurst Bay, Australia was measured at 13 meters (42 feet).

Glossary

atmosphere	A blanket of air surrounding the Earth.
center fix	Finding the center of the storm.
churning	Stirring up into huge, tumbling waves.
clockwise	In the same direction as the hands on a clock.
counterclockwise	In the opposite direction to the hands on a clock.
cut	A journey through the eyewall cloud into the hurricane.
dropsonde	A package of weather instruments that collects information about a hurricane when dropped into a storm.
eye	The calm, clear area in the center of a hurricane.
eyewall	The wall of clouds that surrounds the eye.
landfall	When a hurricane moves over the coast onto dry land.
meteorologist	A person who studies and forecasts the weather.
Northern Hemisphere	The northern half of the Earth, containing the North Pole.
probe	An instrument used for measuring and examining.
radar	A signal that detects the position, size and movement of objects.
radiosonde	A package of instruments attached to a weather balloon and used to measure weather conditions in the upper atmosphere.
satellite	A small spacecraft above the Earth that sends back information about the weather in different parts of the world.
Southern Hemisphere	The southern half of the Earth, containing the South Pole.
storm surge	Sea water lifted by a hurricane and pushed onto the land.
super computers	Very fast computers that can handle a lot of information at the same time.
water vapor	Water that has turned into a gas because it has been heated.
wind	Air moving from one place to another.

Index